Crawl

TARANTULAS

MARTY GITLIN

BLACK
RABBIT
BOOKS

Bolt is published by Black Rabbit Books
P.O. Box 3263, Mankato, Minnesota, 56002.
www.blackrabbitbooks.com
Copyright © 2020 Black Rabbit Books

Marysa Storm, editor; Grant Gould, designer;
Omay Ayres, photo researcher

Names: Gitlin, Marty, author.
Title: Tarantulas / by Marty Gitlin.
Description: Mankato, Minnesota : Black Rabbit Books, [2020] | Series:
Bolt. Crawly creatures | Audience: Age 9-12. | Audience: Grade 4 to 6. |
Includes bibliographical references and index.
Identifiers: LCCN 2018020566 (print) | LCCN 2018021581 (ebook) |
ISBN 9781680728194 (e-book) | ISBN 9781680728132 (library binding) |
ISBN 9781644660249 (paperback)
Subjects: LCSH: Tarantulas–Juvenile literature.
Classification: LCC QL458.42.T5 (ebook) | LCC QL458.42.T5 G58 2020
(print) | DDC 595.4/4-dc23
LC record available at https://lccn.loc.gov/2018020566

Printed in the United States. 1/19

CONTENTS

CHAPTER 1

Meet the TARANTULA

A man on a Texas hillside rests peacefully. Suddenly, he gets a weird feeling. He opens his eyes wide to see a tarantula crawling toward him. It slowly creeps closer on hairy legs. He screams, jumps up, and runs away.

The man had little to worry about. Tarantulas look scary. But these spiders are actually shy and harmless to humans.

COMPARING SIZES

Paloma dwarf tarantula

Mexican red knee tarantula

tiger tarantula

goliath bird-eating tarantula

inches

Big, Scary Spiders

Tarantulas are the largest spiders in the world. There are hundreds of kinds. They often weigh 1 to 3 ounces (28 to 85 grams). Tarantulas are usually brown or black. They all have powerful jaws and fangs. And the spiders are all very, very hairy.

1 to 2 inches (2.5 to 5 centimeters)

4.5 to 5.5 inches (11 to 14 cm)

about 8 inches (20 cm)

about 11 inches (28 cm)

2   4   6   8   10   12

Hairy Legs

Tarantulas have claws at the ends of their legs. They use them to climb. Hairy legs help them move along many surfaces. The hairs let them feel **vibrations** too. These spiders have eight eyes but a poor sense of sight. Vibrations sensed by their hairs tell them when other animals are near.

Tarantula eyes can only sense movement and light and dark.

EYES

TARANTULA FEATURES

HAIR

LEGS

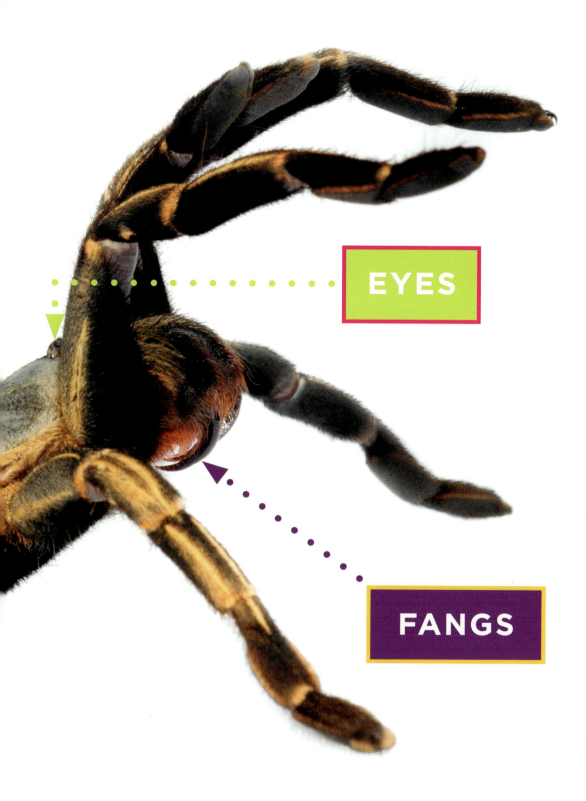

EYES

FANGS

CHAPTER 2

WHERE THEY LIVE
and What They Eat

Tarantulas live all over the world. They crawl around deserts and other warm areas. Many can be found in the forests of South America. Several types of tarantulas live in the southwestern United States.

TARANTULA RANGE MAP

Tarantulas live on every continent except Antarctica.

Living Alone

Many tarantulas live alone in **burrows**. Some use burrows left behind by other animals. Other tarantulas live under rocks and logs. Some even make their homes in trees. These spiders don't make webs out of silk. But they do use silk to make nests.

Where Tarantulas Nest

in burrows

under rocks and logs

in trees

17

Finding Food

Tarantulas come out to eat at night. They wait quietly in the darkness to attack their **prey**. Tarantulas eat mostly insects. But larger types also feed on mice, snakes, small frogs, and toads.

Tarantula fangs release **venom**. It has a **chemical** that turns flesh to mush. The spiders then slurp up their food.

CHAPTER 3

FAMILY LIFE

Male and female tarantulas meet to **mate**. After mating, females wrap their eggs in a silk sac. Females guard the sac. After hatching, young tarantulas grow and molt. During molting, they shed their outer layers. Females look after their young at first. The young spiders then leave after two to three weeks.

Female tarantulas

sometimes eat the males after mating.

MATING

GUARDING THE SAC

MOLTING

Tarantula LIFE CYCLE

Female tarantulas lay eggs and protect them.

Most tarantulas are fully grown after three to 12 years.

Eggs hatch
after six to
nine weeks.

Young tarantulas
molt and grow.

Tarantula Food Chain

This food chain shows what eats tarantulas. It also shows what tarantulas eat.

SNAKES   **LIZARDS**   **BIRDS**

TARANTULAS

INSECTS   **MICE**   **SMALL FROGS AND TOADS**

24

CHAPTER 4

THEIR ROLES
in the World

Tarantulas play important parts in the world. They eat many pests. Quite a few animals eat tarantulas too. The spiders must fight to stay alive. When threatened, they rear up, showing off their fangs. They also flick their barbed hairs at attackers. The hairs **irritate** attackers' eyes, skin, and lungs.

25

Terrific Tarantulas

Humans are a threat to tarantulas too. People who fear these spiders often kill them. **Deforestation** threatens tarantulas that live in trees. They lose their homes. Some tarantulas have become **endangered**.

Tarantulas might be scary. But they're important. They help keep insect populations down. They are food for other animals too. People must make sure these amazing creatures don't disappear.

One type of wasp paralyzes tarantulas. They lay eggs in the spiders. Then the eggs hatch. They eat the still-living spiders.

BY THE NUMBERS

up to 1,000
HOW MANY EGGS FEMALE TARANTULAS CAN LAY AT ONE TIME

ABOUT
25 years
HOW LONG FEMALE TARANTULAS CAN LIVE

1 MONTH
HOW LONG
TARANTULAS CAN
GO WITHOUT EATING
AFTER A BIG MEAL

**1 inch
(2.5 cm)**
LENGTH OF GOLIATH
BIRD-EATING
SPIDER FANGS

**NUMBER OF
TARANTULA SPECIES**

**more than
850**

GLOSSARY

burrow (BUR-oh)—a hole in the ground made by an animal for shelter or protection

chemical (KEH-muh-kuhl)—a substance that can cause a change in another substance

deforestation (dee-FAWR-uh-stay-shun)—the act or result of cutting down or burning all the trees in an area

endangered (in-DAYN-jurd)—close to becoming extinct

irritate (IHR-uh-tayt)—to make painful or sore

mate (MAYT)—to join together to produce young

paralyze (PAR-uh-layz)—to make someone or something unable to move or feel all or part of their body

prey (PRAY)—an animal hunted or killed for food

venom (VEH-num)—a poison made by animals used to kill or injure

vibration (vahy-BREY-shuhn)—a quick motion back and forth

BOOKS

Hely, Patrick. *Tarantulas.* Our Weird Pets. New York: PowerKids Press, 2018.

Keppeler, Jill. *Tarantula vs. Piranha.* Bizarre Beast Battles. New York: Gareth Stevens Publishing, 2019.

Stewart, Amy. *Wicked Bugs: The Meanest, Deadliest, Grossest Bugs on Earth.* Chapel Hill, NC: Algonquin Young Readers, 2017.

WEBSITES

Tarantula
kids.nationalgeographic.com/animals/ tarantula/#tarantula-closeup-hand.jpg

Tarantula
www.ducksters.com/animals/tarantula.php

Tarantula Facts
www.softschools.com/facts/animals/ tarantula_facts/20/

INDEX